Context in Science

Context In Science

Generalists and Root Causes

Kenneth Tingey
PhD
Rex Spendlove
PhD
Miroslaw Manicki
MD MPh

CIMH Global/2020 Program for Global Health
http://2020globalhealth.com
ken.tingey@2020globalhealth.com

Context in Science
Generalists and Root Causes

Kenneth Tingey, Rex Spendlove
and Miroslaw Manicki

ISBN-13: 978-1499222579
ISBN-10: 1499222572

The cover photograph is used under CreateSpace rights.

Table of Contents

Murmurations

Looking to nature for solutions to our problems, we see a remarkable example in the murmurations of starlings. Thousands of them, even more, fly in formation as one organism, responding to phenomena of some kind, often more than one at the same time. We are brought to understand that murmurations are responses to predators and other threats, but this is probably not all, they say. Communal preparations for environmental extremes, such as cold, may factor in.

What is clear and indisputable is that we thus witness the many acting as one, and a very creative one at that. The flow, the composition of the flock as a whole, would be interesting in one animal. Given that it represents the cumulative though of many makes the experience mesmerizing, breathtaking.

Evolution being a fundamental element of all life, there is reputed to be some form of aromaticity to the process of improving our lot. Human will sometimes serves to refute that notion. Famously, we engage in many activities of dubious good. For one thing, we pollute. This is often a mindless profession, where risks to environmental good are assumed for no purpose at all, short of ignorance. In other cases, some thought is carried out, but in highly compromised fashion, with significant conflicts of interest that are unduly persuasive in terms of outcomes. Finally, we see conditions under which wrong-headed actions are carried out as naked manifestations of power and greed. Is it the desired for financial gain or the ability to pull off a scam that motivates such actions? It may be difficult to tell, and likely both factors exist in key instances, but the net result is critically tragic. It may cost us our existence.

With regard to our health, we may lose that as well. Don't look to the U. S. federal government and its scientific priorities for leadership. The organizations in question are stumbling over themselves to predict more and more dire futures and the work to cast shadows on any and all expectations for improvement.

They see not only a less healthy future, they see cascading financial failure, resulting in consistently higher proportions of your income going to sickness. This isn't just growth in costs, this is supercharged growth in costs and in negative outcomes generally.

How can this be? What is it that the starlings know that we cannot fathom? Aren't we learning more about nature each day? What about the stories about the accelerating pace of scientific discovery? Aren't we hearing about how we know so much more from the last decade than from much longer periods before?

The problem isn't in the lack of knowledge, but in its application. For one thing, much research is simply unknown, lost in the forest of knowledge. This in part is due to the structure of how science is conducted. Scientists are rewarded for coming up with new ideas, but not in their application. Most scientists are thus specialists. They come up with ideas. They test them out, typically by validating that they do not occur by chance. Then they publish the material in the form of a document once other scientists, "peers", agree with their conclusions.

There is a joke in the U.S. research community that there are three phases in science. Phase one, is idea generation, a creative stage where an idea is identified, often noted for its absence in the scientific literature of associated subjects. Phase two is a prototype phase, where the idea is proven out, including the construction and demonstration of a prototype, an example of the idea being applied. This may include the actual construction of a physical idea, or it may include a worksheet or questionnaire that lays out the idea so that it can be understood and tested.

The third phase, implementation, is where the idea typically "falls off the table". There are more reasons for this than can be adequately considered at this time, but it is safe to say that they make acceptance of the new idea very problematic, indeed.

Wrapped up into three, the implementation problem is difficult because of ignorance of the benefits that could be brought on by the idea, inadequate available resources to maximize on its effectiveness, and resistance by parties whose interests may be compromised by introduction of the idea.

If the first problem is not solved, the other two don't really matter. It is the first problem that we consider herein.

With regard to various aspects of health, nutrition being one of them, we often stumble. The Institute of Medicine recently published a pessimistic view of this phenomenon, indicating that we are doomed to live "shorter less healthy lives" than people in other parts of the world. How can this be, with the many advantages that we have had? To be sure, we are not behaving as the starlings with regard to knowledge, with regard to newly found root causes of both healthy conditions and sickness. We are not elegantly cooperating in murmurations of life and health, we are slamming into the barn door.

Chronic diseases? They are out of control. Poor lifestyles? They are preferred above all else. In this, our wonderful educations are surely deficient if they do not lead us to learn of and to embrace better options. We need to be led by the hand – together and individually. Scientific generalists, empowered by expressive knowledge models that they design and implement themselves. Perhaps our thought can be reoriented toward optimal living patterns. Hopefully, we make it before other, pernicious natural causes take effect.

The generalist function

Classical economics has been criticized for its reliance on what has been referred to as the "invisible hand". The invisible hand phenomenon was held to be an automatic leveling mechanism that brought balance to a market. In a sense, the invisible hand phenomenon was held to result from large markets in particular. The larger the market, the more there was a chance that balance would be achieved, through the actions of individual actors, each working to achieve personal objectives. Such a system would optimize itself, indeed, it would come to a state of balance as a natural process.

Ah, such a noble era, with the postwar steamroller of goodness emanating from North America, in part allowing for the germination of such naïve, even self-serving thoughts. Presumed as an attachment or an adornment of that philosophy was the assumption that the invisible hand would naturally guide itself in the direction of such goodness. As Americans and as beneficiaries of the actions of many who sacrificed much, if not all, in favor of such goodness, we also acknowledge that this association was defined by choices, many carried out over generations, to elect higher paths, not from any inherent superiority. It is in the ongoing selection of better options that this heritage lies. Just as Jesus of Nazareth said that he could raise "sons of Abraham" from the rocks, we need to make such choices carefully and thoughtfully, lest we, too, go down the path to tyranny and waste.

This we have done in many cases. Our industrial foundation is littered with wasteful projects where better options exist. Lest we unnecessarily offend, we will not rehearse these. In the parlance of the Wild West, "you know who you are".

Someone needs to piece the science together, a constant occupation. As mentioned earlier, we do not believe that it is going to happen automatically, at least not in the course of the next few years. Reason must be applied to the general body of knowledge. Power relationships, political, legal, and commercial, need to be arranged and rearranged to tilt the balance in favor of ongoing health and prosperity.

To bring this about, there needs to be a concerted effort to scan the horizon for applicable developments, to be on constant watch for root causes of unwanted conditions and of natural groundings for health. Of course, in this we mean that there needs to be a commitment to understand the fundamental requirements of good health and well-being as well as disease conditions. In this, we refer to the causative factors of well-being at the most fundamental and elemental levels possible.

In the search for knowledge, seeking out root causes is a lot of fun, to be sure. This is the tip of the sword with regard to human knowledge and progress. Such questions deal with the physical sciences to be sure; understanding of physics, basic chemistry, both inorganic and organic, and the study of life's foundations are of critical importance. These being the fundamental pillars of life, they must be understood as prerequisite aspects for policy and understanding.

As we like to repeat, although nature does not sit around the table when policies are negotiated, she *will* have the last word. To the degree that policy misses the mark with regard to such root elements of science, undesirable outcomes will occur. The job of the generalist is to study out and identify such conditions with regard to our understanding of nature. Awash with the knowledge of ongoing research efforts, they are in position to direct resources and attention to where they belong.

This is not to say that science does not or cannot apply to social and economic affairs. On the contrary, one reason we need to gather and interpret large stores of data is to make scientifically-grounded interpretations that will hold true in social, political, and economic contexts.

There is a never-ending debate with regard to fundamental research forms. To what extent should resources be directed at basic research as opposed to applied science? Such questions need to be addressed with the root cause assumption, as research is highly subject to social and economic factors, apart from whether it is considered to be "basic" or "applied" in nature.

What if, in the current environment, esteemed scientists were to come up with palliative approaches to the needs of persons with late stage smallpox, given that few people now suffer from that condition? What if some such research groups were to establish means of extending the lives and improving mobility, if only slightly, of persons with advanced polio? How would such developments be evaluated, given more pressing needs elsewhere?

These would surely be considered beneficial developments by all. Any progress in the alleviation of pain and suffering and the reversal of disease conditions needs to be considered as progress.

On the other hand, one has to ask, now that such conditions are rare, perhaps there are higher priorities for research and implementation efforts. Consideration of current ills certainly would be a priority, but also considering eliminating root causes as object.

We might say that priorities should be established keeping current economic questions in mind. Such factors should enter into the mind of generalists as they review scientific literatures. There are those who would say that we would need to be careful to protect existing economic commitments first, and to consider possible benefits later. After all, the economy demands full employment as an article of faith and a new development may result in a net loss of jobs, political poison.

In this light, how is some "ivory tower" generalist scientist, sitting around reading journals going to understand such things, to grant sufficient weight to such issues?

Truly, such awareness on the part of such a scientist may be unlikely, but a part of the generalist task is to inform himself of what needs to be understood in the course of his or her review of the research and of social and economic trends. In a sense, the questions of scientific and social applicability may be incompatible or antagonistic if brought under consideration at one time; concern for issues in one realm may curtail prospects in the other. On the other, the challenge of the generalist is to identify the opportunities as well as the problems so that attention can be focused on the issues and decisions can be made, including the commitment of resources to improved use by means of the findings.

With deep understanding of the science, the generalist to be sure should "push the envelope" even as he or she considers the work of specialists, who themselves need to be free to do the same, within ethical and legal bounds. This is what the academic tenure system should be about, although it clearly often is not, serving other purposes, counter to the pubic good. Economic and social issues must be considered in their turn, adapting in their turn to both positive and negative economic measures, but the science at this stage need to prevail, particularly the science of root causes that can free us from our bonds.

Peer review, fluidity, and the public good

So the generalist sees interesting, new developments that have passed "peer review" and have been published. To what degree should peer review, particularly as understood by networks of specialists, be viewed as the gateway to introduction of new developments in fact?

In an environment of fluidity, the peer review process comes front and center. Fluidity is the flow of knowledge among people and technologies. In the methods-based management approach, fluidity is analogous to the concept of cognitive legitimacy.[1] An organization with high levels of cognitive legitimacy in short does what it is supposed to. This is to say that it is an effective vehicle for accomplishing the objectives it is generally understood to be carrying out. In business management terms it thus could be said that an organization with a high degree of cognitive legitimacy is fulfilling its mission.

Any arbitrary limitations or barriers to fluidity or cognitive legitimacy would thus hamper the organization in its ability to fulfill its mission. With regard to fluidity, then, another term for cognitive legitimacy, any limitation to the flow of scientific knowledge as represented by peer reviewed science, calls into question the legitimacy of the systems of science overall. It is generally understood, or at least, hoped by the general public that valid scientific findings work their way into general use.

Unfortunately, many now question whether this is the case. Even more unfortunately, there are many that are glad that science is not more readily used. Trust is low; many non-scientists discount what is presented as scientific advancement on political grounds, thinking that scientists have political agendas. No doubt many of them do. Much scientific work has been criticized as being "socially constructed", an interesting development in which social

---

1   Tingey, K. B. 2009. *Methods-based management.* San Diego, CA: University Readers.

scientists appoint themselves as evaluators of physical scientists. The criticism here is that scientists set up the rules and "stack the deck" in favor of their preferred solutions. Indeed, this can be considered a pervasive human trait, a reason for established longstanding means of evaluating and balancing out the outcomes.

Of course, the scientific method's purpose is to ferret out such issues; much work has been done, dating to ancient times, to establish credible scientific findings. The generalist's function is critical to such an effort and to establish a pervasive means of gathering and evaluating appropriate stores and flows of data.

Does the public deserve a non-watered down, non-engineered form of scientific truth? What would the public want in this regard? Clearly the public is comfortable with some level of denial. Many considered such preferences in light of their personal freedom. Scott Adams, the author of the Dilbert cartoons, has authored a book "When Did Ignorance Become a Point of View?"[2] The answer, of course, is when democracy was invented. Denial is a long-standing tradition.

The *2020 Program for Global Health* and its objectives with regard to chronic disease eradication in particular are based on the presumption that such scientific relativism has two limiting factors, at least with regard to public and private health: pain and death. This is to say that even though political will may fall short of full concordance with the realities of nature, there is a strong mandate for reducing pain and eliminating death where possible, particularly painful, premature, debilitating high mortality conditions.

The theory behind this is that these two conditions touch all of us, wealthy and poor, social and political insiders and extralegal orders. They are universally unwanted conditions; people can be assumed to want to eliminate both, or at least put them off, as

---

2   Adams, S. 2001. *When did ignorance become a point of view?* Kansas City, MO: Andrews McMeel Publishing.

long as possible. This is not to say that some number of individuals within the population may not be suicidal and otherwise act against their interests in terms of pain and death, but that society as a whole, parents and social and political leaders, universally wish for an end to unnecessary physical and mental forms of pain and premature, untoward causes of death.

The authority of generalists is grounded in such presumptions. We all want solutions. The generalist scours primary sources of scientific findings looking for them. As such, they need to develop an accurate understanding of the way things are conducted currently in health and other affairs and can thus identify potential improvements, with some understanding of their scope. They need have the resources and capacity to identify teams of specialists and to help to call attention to their work.

We all benefit from and depend on the success of generalists involved in these critical functions.

Generalists with the "capital G"

The function we describe here is that of a fully-empowered generalist. A generalist can be distinguished from a specialist in terms of the scope of his or her efforts and interests. To put it succinctly, whereas a specialist will often be spotted tinkering in the laboratory, a generalist, well-known for laboratory work of his or her own, will more likely to be found reading. Whereas a specialist or a team of specialists will associate with others with similar focused interests, generalists will have wider ranging professional circles, typically marked by association with researchers they become aware of through their readings and other studies who represent areas of study they find to be of particular interest.

Rex Spendlove's experience is particularly useful on this score. As founder of *HyClone Laboratories* in the 1970s, he took it upon himself to develop a broad understanding of cell culture technologies and and research in related fields. As a result, he methodically, regularly studied all relevant works. To do this, he reviewed all issues of *Current Contents*, the best comprehensive, timely source of information about major peer-reviewed scientific journals available at the time.[3] Carrying current copies of the index virtually *everywhere he went*, he marked interesting articles based on abstract descriptions. A staff member would regularly travel to the nearest medical school library, about two hours away, to obtain copies of the articles themselves, all of which he reviewed and filed away for future reference.

For years, he built his personal and corporate library in this way. He expanded his review efforts to include all weekly issues of the journals *Nature* and *Science*, the journal *Microbe* by the American Society for Microbiology and *Genetic Engineering & Biotechnology News*, which has shown to be a reliable biweekly trade journal.

---

3   Several editions of Current Contents continue to be published by
     Thomson's *ISI Web of Knowledge*.

With time, Dr. Spendlove learned that his firm enjoyed a preeminent position in the industry in terms of knowledge and perspective. Knowledge ran deep in the company's principal product areas, also wide due to the constant reading and review efforts. Personally, over the years he developed a perspective that could have been achieved in no other way. No "Johnny come lately" to any of the fields of study in his reading regimen, he became a generalist of the kind we describe herein.

To use a common metaphor, specialist scientists study "trees" while generalist scientists study "forests". This is a figurative, not a literal comparison. Hopefully it helps to make an important distinction.[4] Ultimately, we need to understand generalists in their ability to scan the horizon of scientific study and find not only the best work in related areas, but the most relevant work in related areas. In understanding root causes, relevance is of more than passing importance. In our day, perhaps the best example of this is in cancer. It is far more important that we never contract the disease in the first place than that we have elegant means of dealing with hopeless cases late in life. One problem we face in our time is that the former statement can be considered controversial in any way. Unfortunately, it can and it is.

So, an effective generalist in this case is a person who can evaluate all that exists in related fields, judging as to the validity of the works in question as well as to their association with root causes. It is also important that the generalists in question provide guidance as to how that work can be adopted and expanded on.

Generalists apply a form of rigor of their own. Of course, there are specific methods applied by generalists, ways of evaluating research using statistical and scientific methods. Specialists can also write reviews of related works. They engage in what is called

---

4   In an environment of fluidity, where expressive knowledge is a guiding factor, everyone is deeply involved in trees, in their design and use. Generalists engage themselves in the design of trees that link together those of specialists and others. In the context of their responsibilities, and only there, they are in position to link the other trees, or models, together.

meta analysis, which is a way of combining the results of individual studies to compare and to weigh their various implications.

Specialists put the bulk of their time and the majority of their energy into specific areas of study. To survive and thrive, they must be more and more directed and detailed about their work. Thus must experiment. They must observe. They must drill down to the most fundamental elements of their areas of study.

This is contrasted with the objectives of generalists, who are on constant watch for new and better ways of carrying out the life's activities, regardless of their nature. As understanding and committed as generalists may be to their initial areas of specialization and knowledge, generalists must be willing to turn it over in a day, were they to be convinced of a better way, of more fundamental levels of knowledge, of more natural and dependable outcomes.

For this and for other other reasons, specialists are often mistrusting of generalists, if not altogether dismissive of them. Specialists typically criticize the rigor of the work of generalists if they can't get away with ignoring them altogether. Given that generalists may not be experienced in laboratory work, this, too provides fodder for criticism. Finally, specialists criticize generalists "for not understanding the math" inherent in their work, a criticism that may or may not be valid.

Laboratory work is of critical importance, central as it is to the discovery of phenomena and validation of what we think we know. Laboratories, of course, cover a vast array of environments. In fact, any situation or set of condition can be considered laboratories. Natural environments make for the best laboratories, but in some cases, contrived conditions are necessary. A laboratory could be a classroom or a workroom as much as a biological wet laboratory with specialized instrumentation and highly controlled conditions.

Laboratories are populated with people in the act of studying something. Truly, the scientists in question may not actually be "in" the environment under study, but they are clearly looking at some natural phenomenon, if not a contrived situation that they hope is similar enough to nature to provide some insight as to how nature behaves.

Lifestyle

The lifestyle of the generalist is the lifestyle of a reader. It isn't that a generalist ascribes to knowing everything about everything, but the object is to know as much about some defined area of knowledge as can be known. There is no available alternative to that of constant, thoughtful, indeed, voracious reading in achieving such a goal.

Regardless of the subjects in question, generalists will always have stacks of reading. There will be certain key journals that must be absorbed continually, whether on a weekly or monthly cycle. This "corpus" of published material needs to be selected carefully and methodically. As a scientific practice, the selection process is a sampling procedure and must be considered as such. It must be representative. It must be comprehensive. As with all forms of scientific data stores, it must be unbiased.

With time, the nature of the review task becomes more and more comprehensive. The committed generalist will come to know the body of work in key areas, not to the depth of the specialist that lives in the midst of related research activity and its related sociality, but with more breadth.

The generalist function involves the constant collection of literature reviews and other combined forms of study. Where data is brought together and statistically evaluated and studied, the reviews are referred to as "meta analysis". Collection, acquiring, and reading of associated materials, the items making up relevant bibliographies, it fundamental to the generalist's task. Generalists are constantly stitching together the work of varied research groups. Not only organized sampling techniques need to be used, but meta analysis of many studies is required.

From an *Optimum Performance Living (OPL)* perspective, as proposed in the *2020 Program for Global Health*, the generalist need be constantly looking for root causes – both for the good and for the bad – in order to resolve issues and introduce dependable grounds for health and well-being.[5]

Coding and classification of knowledge elements is an important habit, a task that leads to the development of models and sketches that enhance understanding and help to clarify the context of relevant conditions. These can be sketched by hand, something of a useful artistic talent. Such models, referred to in education as "advance organizers", are very useful in understanding relationships and helping others to do the same. Coding and classification are primary elements in the development of expressive knowledge forms and are considered in great detail in the methods-based program.[6]

What is missing in general is a way of directly computerizing and applying such knowledge to the practice. Presumably, such a task is taken over by computer design specialists who take the knowledge product and convert it to useful forms. Our point is that this does not work in fact.[7] The result is incomplete solutions, hemorrhaging of costs in all respects, and dis-empowerment of knowledgeable people, generalists among them.

As Howard Hughes indicated, he always kept up on the "trade journals", this aspect of the knowledge must also be continually read and understood. Generalists need to be aware of what is going on "in the street" in related areas.

---

5   Farnes, L., Manicki, M., Lee, S., Spendlove, R., Ostojic, D., and Tingey, K. 2014. *Optimum performance living: We can have it if we want it.* Logan UT/Warsaw, Poland: 2020 Program for Global Health.
6   Tingey, 2009.
7   Tingey, K. B. 2014. *The solution: Permanescence.* Logan, UT: Profundities LLC.

Of course, there is much of the specialist in the generalist. In most cases, it was when scanning the horizon for an area to study that caused the specialist to make commitments to a specific field of study in the first place. Understanding of relevance is important to such decisions. Once the commitment is made, however, the specialist becomes a cheerleader for the field in question. This is a good thing, inevitable and necessary. All perspectives are in need of advocacy and support. This can be a counterproductive, even pernicious development, however, in cases where root cause elements render the area of study less interesting.

Of course, one of government's main tasks in a dual control environment is to make resources available in support of scientific and research inquiry generally to take much of the risk out of the role of specialization. Society needs to provide "golden business deal gone sour. Should an area of study be less critical to the health and well-being of the general public as per current parachutes" to specialists much more than to business executives who find themselves on the losing end of transactions. A scientific "bet", well-placed, is far more valuable to society than the typical conditions, that does not mean that it constitutes a waste of resources and of time. It may contribute, for example, to the general understanding of science and of nature. It may also provide groundings for additional research that will itself open new doors even higher up the chain of root causes.

There needs to be a way of evaluating, of making meritorious judgments between more than one competing alternative. This needs to be done in a way that not only makes meritorious choices, but that results in conclusive outcomes. This is where generalists come in, of course. Working with policymakers and with other relevant parties, advocates and the public, generalists need to be able to document if not activate solutions that bring desired findings to general use.

Tools

What, then, are the tools of the generalist? A first rate education is necessary, of course. This needn't include scholastic preparation in everything, but an education with defensible rigor in areas of interest and relevance to the area in question.

There are those that believe effective generalists can only come from areas of specialization underlying subjects at hand. That indeed may be true in certain fields, particularly where extensive knowledge of mathematics, physics, and the hard sciences are involved. This is surely the case, as emphasized by Woodger in areas of biological study. By definition, as he indicates, good science stretches beyond the limits of spoken language.

> … biology, like other sciences, begins with observations on the familiar things of everyday life, and its results are at first adequately expressed in the language of everyday life. But as it advances it widens the scope of its observations. The invention of successful hypotheses brings with it the invention of new apparatus for observation. The records of observations made with such apparatus are meaningless apart from the theory of the apparatus – the hypothesis which have led to its invention. Is it not to be expected then that sooner or later the linguistic habits of everyday life will cease to do justice to this increasing complexity and novelty of observation and hypotheses?[8]

This being said, many generalists make use of commentary from fields other than one's own. This is a necessary factor, as degrees of separation between investigators and application of their knowledge are necessary. For one thing, there are embedded conflicts of interest in the process. Specialists, having committed

---

8   Woodger, J. H., 1952. *Biology and language: An introduction to the methodology of the biological sciences including medicine.* Cambridge, UK: Cambridge University Press.

all to dedicated fields of study, understandably want their work to count for something. They will always seel hard. Generalists need to be able to function responsibly and authoritatively in such an environment.

Talent, skill, and initiative can prepare a person to serve in the generalist role even without special knowledge as it applies to related fields of study. At times, such factors help such people gain credence from the public and respect and acknowledgment from specialists and other leaders in related fields. Legitimacy, however, comes from a balanced assessment. Indeed, generalists need to have the strength of understanding and functional legitimacy sufficient to steer research as well as practice and commerce away from areas of specialization that lack relevance with regard to root causes or that do not have research products that stand up to acceptable levels.

There is something inherently entrepreneurial in the generalist role. Many generalists are self-selected and self-appointed. A generalist must be confident, to be sure.

A generalist must be multilingual, including a healthy mix of tacit, explicit, and expressive languages and knowledge forms. Skill is not only needed with regard to the languages of science and those of the people, but also with regard to fluidity, the ability to define and deploy associated processes. They must be able to understand processes presented to them in generative, expressive forms and they must be able to understand extant models and systems. In this way, they are able to position relevant works within the system so that it is recognized and used in appropriate contexts. They are the ones, ultimately, who control Castell's switches, the "privileged instruments of power" on the network and in society.[9]

---

9   Castells, M. 1996. *The rise of the network society.* Volume 1: The
      Information Age: Economy, Society, and Culture. Malden, MA: Blackwell
      Publishers, 471.

Generalists must have the last word. They must be thus empowered and trusted, and they must have the capacity to accurately judge between realities and possibilities.

Facility in language, as Woodger indicates, is a challenging prospect. On the one hand, a generalist must bridge the language gaps embedded in the various literatures in question and spoken languages, but must be particularly cautious so as to not fall prey to metaphors available to spoken language.

> Metaphors ... are often suggestive and may be harmless enough if they are recognized for what they are. But at best they are makeshifts and substitutes for genuine biological statements, and the fact that recourse is had to them is surely a sign of immaturity. Science demands great linguistic austerity and discipline, and the canons of good style in scientific writing are different from those in other kinds of literature.[10]

Indeed, the reward for use of metaphors can be public popularity. The masses, including policymakers and commercial purveyors, can thus take from statements what they will. As can be readily seen in our time, linguistic imprecision allows words and phrases to be stretched to the far reaches, rendering context and validity totally absent.

This can serve to invalidate findings where tacit and explicit forms of knowledge hold sway, but expressive forms of knowledge can be used to enforce much higher levels of context and meaning, resulting in higher levels of scientific precision. Truly, more than a few critical knowledge elements have been lost in a fog of metaphor when a scientist reverts to rhetoric to too great a degree.

This makes sense from one perspective; the answer to validity lies in an understanding of specialized, detailed underpinnings of

---

10 Ibid., 8.

knowledge. The main problem is that relevance cannot thus be judged. Independent perspective is necessary. This is the primary function of the generalist.

The human capacity for pattern recognition is well-established. Generalists need to be expert in this. They do not need to have the depth of understanding of specialists in every case. An understanding of inputs and outputs may suffice. Understanding, if not trust, that specialists do understand the complex underpinnings of the subjects at hand is an essential aspect of achieving Optimum Performance Living (OPL) and providing conclusive, permanent solutions.[11]

The generalist is not the ultimate judge, to be sure. As dual control is dependent on the ongoing collection and evaluation of data,[12] validity of a chosen path can be determined independently in a number of ways. It is through ongoing, thorough evaluation of data that efficacy is to be determined. For validity and reliability to be declared, and for comparability to be enjoyed, design, management, and control of the details of underlying processes with full, deep understanding of the phenomena in each case are essential. Generalists play critical roles to be sure – in some cases creating such processes, weighing and prioritizing the work of specialists so that in appropriate contexts, their work gets considered, if not used.

Using the music performance model as an example, both generalists and specialists engage in composition, users of the system being the performers. The composition is a collaborative one, where generalists provide integrative models and specialists

11 Tingey, K. B. 2013. Chapter 2: Seeking truth from facts. In *The angels are in the details: Control and regulation ... in a good way*. Logan, UT: Spendlove Medical Research Institute, Profundities LLC, 2020 Program for Global Health, 26-56.
12 Manicki, M., Tingey, K., Farnes, L., and Ostojic, D. 2014. *Dual control or certain derailment*. Logan, UT/Warsaw, Poland: 2020 Program for Global Health; Manicki, M., Putri, A., Ostojic, D., and Tingey, K. 2014. *The big step forward: To knowledge-driven universal coverage*. Logan, UT/Warsaw, Poland: 2020 Program for Global Health.

detailed protocols and actions. This is a very human endeavor; priorities must be set and alternatives weighed, mostly by the generalists. Given a particular context, a particular problem set, however, it is ultimately the skill and knowledge of the specialist that holds sway.

There is a perceived, great divide between the physical sciences and social science, but there needn't be. Much more is known about the physical and physiological underpinnings of life that effect, if not drive behavior than we not use. Generalists, armed with expressive knowledge tools and concepts and positioned to make use of them, could provide access to all of us of the fruits of such knowledge when they can be best applied. This will allow for more connections between physical and behavioral realms.

Ultimately, this is where generalists can best serve. Prepared to function in this netherworld between fact and opinion, they can work together to validate and activate decision processes in the contexts where they apply. They must be able to arbitrate as well. In an environment in which fluidity is present and dual control is achieved generally, their decisions hold sway, allowing them to exercise their skills and be accountable to the people and to their collaborators.

The friendly generalist

A good generalist will tend to know a lot of people. He or she would have read the work of many people. Given their wide range of interests, good generalists tend to know of many scientists and others in a broad array of activities. An effective generalist will need to have good interpersonal skills and be persuasive.

Standard protocol in the scientific community is to correspond, to meet, and to share ideas and learn from one another. Generalists will tend to "cut a wide swath" in this way, participating in a wider circle of acquaintances than would be typical of specialists.

Of course, there are generalist-oriented associations of scientists as well as those that apply mostly to areas of specialization. Such associations are beneficial to be sure, but they should not replace direct relations with specialists. Without such ongoing contacts, particularly those driven by growing awareness of new solutions to root cause problems, a form of "group think" can easily settle in, undermining any efforts to achieve root cause-driven "murmurations" by society.

The collaborative generalist

Speaking of trees, as can be seen in the Figure *Groundings for Optimum Performance Living* on the next page, the generalist task is best served when matched with a payment structure that is both unitary and authoritative.

When considering fundamental causes rooted in the driving functions from below, from the roots, generalists match them with policies and methods acceptable to and legislated by authorities in question. In a dual control environment, this includes all relevant authorities and functions, whether science-based or not.

In such cases, it is critical that fluidity exist in a systems environment to support each decision in its rich, individualized context. Thus, as recently outlined by Dr. Forest, sound science can be matched with policies and preferences within society, in all cases supported by available resources, given limits.

> Each year we spend more on health, and each year we ask ourselves how this money could be better spent. On new pharmaceuticals? Better paid physicians? New hospital beds? Better prevention? More research? We do not even know who should be answering these questions: our political leaders? Experts? Citizens themselves?[13]

Dr. Forest indicates that we need a "new language" in support of a "new synthesis". This is that. The synthesis needs to be a true one.

---

13 Pierre-Gerlier Forest. International Journal of Health Policy and Management [2014, 2(2), pp 55–57],

Far too often what is called synthesis is simply the loudest voice in the room – if not the only voice in a remote enclave.

Figure: Groundings for Optimum Performance Living

The *2020 Program for Global Health*, committed to knowledge-drive universal coverage, subscribes to the idea that each person deserves to partake of the fruit of the tree of optimum performance living, knowing that they are being guided in the most beneficial direction possible, given their desires and personal preferences.

For this to occur, "somebody" needs to make decisions with regard to cellular biology that understands the cellular biology, the "-omics" that understands them, the physics, the pathology, the immunology, and all other relevant areas. That "somebody" needs to work in tandem with valued partners to piece together the work product of skilled specialists, to encourage the kinds of decisions Dr. Forest makes reference to, to serve a valued function to society. We need people in such roles to be our "Beethovens" of health and medicine.

It is important to note that a new language and a new synthesis are necessary in order to even understand what we mean. In an environment of fluidity, where there is understanding of and commitment to the wide use of expressive knowledge forms, where there is country-level commitment to dual control and to the administration of policy in a data-driven environment with clearly-supported generalist and specialist roles, such goals are very achievable. Where there is an environment driven by methods and data, the murmurations will occur.

Conversely, such goals cannot be met where there is faith in and dependence on an "invisible hand" in health and medicine. It cannot happen with machines that think or by convening specialists in a big room. Although their technology is of critical importance to success, the Silicon Valley crowd, with their command-and-control mentality, will not bring us the solution. Generalists and specialists, armed and ready, will.

Actually doing something

In terms of a system of support and in terms of fluidity, generalists are in position to assist in the organization of key semantic structures in the system. The context of new findings needs to be appropriately defined, ready to be activated by quantitative and qualitative data made available through the system. Generalists have a specific role of arbiter of such semantic knowledge forms, making use of expressive knowledge forms in particular to assure that appropriate integrative processes are followed.

Expressive knowledge is a means of organizing knowledge that goes beyond writing documents and creating other static forms of knowledge. Expressive knowledge involves the expansion of such works however, which are considered "explicit" as opposed to "expressive" forms of knowledge.

We are proponents of a system underlying the validation and presentation of science that includes expressive as well as explicit knowledge forms. This is to say that publications should not only include written presentation of the issue and associated findings, a logical model as represented by that research should also be included using parsimonious, expressive knowledge forms. Such additions to publications would thus make use of Aristotle's hypothetical syllogism structure to demonstrate and extend the proposition in question.

Use of such extensions needn't involve complex, nor confusing models, but the simplest of approaches to tree design, based on five straightforward concepts that are both intuitive and powerful. Using the five concepts, propositions can be computerized and extended to the masses in highly specific circumstances, following the branching of many such steps.

Who needs science when we have "kumbaya"?

Related to faith in the invisible hand is a hope that we can resolve the grand challenges of mankind by somehow "getting along" better.[14] From this frame of reference, we do not need a systematic approach to organizing, prioritizing, and utilizing knowledge. We simply need to get along better. We need to get together more often. We need to be nicer to each other. This is a common theme in professional confessionals and soft exposes.

Dr. Otis Brawley of the American Cancer Society describes bad behavior on the part of physicians. Few would disagree. The point isn't simply *that* it happens. It is that it *can* happen. Any system based on the assumption of virtue is tenuous to be sure.

> Ego, arrogance, and excessive self-confidence lead doctors to confuse what they know with what they believe. This confluence of forces leads to a closed mind and has led to advocacy of many interventions that, after years of use, were ultimately found to be not useful or even harmful.
>
> I have met my share of doctors who would have been great snake-oil salesmen. I am amazed by the number of medical leaders who seek power and prestige through habitual lying.
>
> Health-care providers, hospitals, drug and device manufacturers, and insurers all need to truly focus on the best interest of the patient. Some try to do the right thing, but some hospitals become more consumed with competition than with providing good care. Some drug companies seize control of medical education and advance their products through medical

---

14 "Kumbaya" is a song often sung "around a campfire" that is generally held to support a feeling of cooperation and togetherness.

policies, perverse incentives, and thinly disguised (or blatant) bribery.[15]

"Stop it!" is the message.

Dr. Brawley admits, although you have to piece elements of the confessional together from his book, that there are enormous gaps, even centuries-long, between discovery and use of the products of research. In the case of prostate cancer, for example, current practice dates to the 1840 in spite of many subsequent innovations that lie unused.[16] He specifically holds "medical subspecialists" to be responsible for "practice guidelines" that bring about "almost every medical travesty".[*] As he says in a particular case:

> These doctors betrayed the oath and fundamental principle of medicine: *primum non nocere* – first, do no harm. Their actions warrant double contempt because they benefited financially from doing harm.[17]

The fact that he says it doesn't really make it any better. The problem is that they have the option of making such wrong choices in the first place. What we are talking about here is akin to the symphonic model. Options available to symphonic performers depend on the context of musical notation laid out in advance by the greatest minds in the world, in our time, and in the case of the musical Masters, at any time.

Nonetheless, it is the *behavior* of the physicians that Dr. Brawley criticizes, not the environment in which they function. In the journal *Science*, there was a recent editorial "Meeting Global Challenges" that similarly struggled with the issue, but without

---

15 Brawley, O. W. 2011. *How we do harm: A doctor breaks ranks about being sick in America.* New York: St. Martin's Press, 282-283.
16 Ibid., 237.
*   It would be very nice if he had called for a systematic program to identify and reinforce generalists, particularly in the interpretation of available science.
17 Ibid., 241.

calling attention to generalist/specialist issues or further definition, support, and empowerment of the generalist function.[18]

The result from bringing specialists together is a cacophony of voices, of selective interests arguing their specialized causes in zero-sum entanglements that will resolve themselves at best through rhetoric. Such assemblies fall prey to the conundrum of the *Tower of Babel*, translational puzzles that cannot be resolved in the aggregate. As Woodger warned, not only is the temptation present in such venues to be linguistically fluent, that indeed is all that there is.

In this we see physical science struggling with social science issues. The awkwardness of the moment would induce humor if it the stakes weren't so high. The irony is in seeing towering geniuses from the fields of physics and mathematics and chemistry and biology struggling with the same issues as those corner drug stores, local country clubs, and generic public or private organizations. Such natural scientists typically don't believe in the validity of social science as an enterprise, but neither do they know how to correct resulting defects, even for their own purposes. Thus, science and its uses fall prey to the ills of the country club, at once subject to the autocrat and to prerogatives along the line.

One result is a disassociated atomization of science. Many islands of specialists inhabit the famous "silos" of the academy and industry. This surely results in duplication of efforts. What there is *not* is an "invisible hand" that the economists similarly miss.

What is to be done? Given our propensity to call for "root causes" generally, we should do so here. Are there any rigorous, scientific groundings for behavior, particularly behavior among groups leading to beneficial, if not optimal behavior? There are those that would say that this is what evolution is. That would be an acceptable definition from a natural science perspective. Such

---

18 Sharp, P. A., and Leshner, A. I. 2014. Meeting global challenges. *Science*, 343, 579.

studies would thus be warranted. From a literary standpoint, this is quite common – calls for survival of the fittest and such.

From the standpoint of rigorous scientific pursuit, however, there is a vast gap between ecological studies of populations and behavioral studies of people in groups. Dr. Tingey of our team has long thought that schools of business and of organizations would be best positioned within natural science, being guided by ecological science administrations, where there is knowledge of and rigor toward the study of groups beyond the "next big thing" conundrum of the business schools, which is costly beyond knowing and an embarrassing menace to society.[19]

We believe that there are groundings for such efforts, not only in theory, but with practices covering the last hundred years. Such efforts, though well-known, with measurable results, have not been fully-appreciated. Certainly their potential contribution has not been acknowledged. Details of this have been documented under the rubric "methods-based management", which we hold to be the sole inheritor of the scientific method with regard to

---

19 Interestingly, an interesting update on the "next big thing" comes from the traditional font of such work, the *Harvard Business Review*. In May 2014, Julian Birkinshaw, in "The big idea: Beware the next big thing", seems to unwittingly call attention to the problem by stating, "Where do management fads come from? A few emerge fully formed from the minds of academics and consultants, but a vast majority come from corporate executives experimenting with new ideas in their own organizations", 6=52. Which is worse, an idea thought up by an "ivory tower academic" or another thought up by someone in the trenches, in private employment whose principal commitments likely skew the results? What ever happened to the scientific process? We find ourselves "on the other side of the river" on that count. Notably, *HBR* itself has been the publisher if many, if not most NBI's.
In a March 2014 article in *Army Sustainment*, "Management fads: Beware of the next big thing" (4), Christopher Paparone and George Topic describe a recycling of similar ideas under new names. Specifically, they call attention to "total quality management" as having roots in earlier movements as described in the Methods-Based Management approach, but much more narrowly.

management.[20] It is this approach that is situated on the scientific method, being data-driven and responsive to constant application of scientific thinking in "real world" environments.

Information management is central to success in this matter. Certainly this is true of nature, biology in particular. As underscored by Woodger's writing, biology can be considered the study of information processing within nature, given pervasiveness of the information sharing with regard to living things. Dr. Tingey has published an essay on that subject, demonstrating the use of genetics-based information models in an enterprise context.[21] It is an approach with fundamental evolutionary groundings.

How best to achieve permanent solutions? Once again, we look to musicians as the experts in the application of scientifically-grounded information solutions. Composers, generalists to the core, famously went to the mountains to create solutions. Often considered irascible if not antisocial, they needed solace and time to think through their creations, in the absence of distractions and compromising interactions.

Ultimately, murmuration as to optimum performance living represents achievement of these goals in timely, efficient ways. Closing the gaps between nature, understanding, and behavior is an appropriate goal for our times. While overdue, and while a difficult endeavor to achieve, it will constitute a monumental

---

20 Tingey, K. B. 2009. *Methods-based management: Breakthrough performance on leaner budgets.* San Diego: University Readers.

21 Tingey, K. B. 2013. Chapter 3 – Molecular groundings for the logic of organizations. In Kenneth B.Tingey, *The angels are in the details: Control and regulation ... in a good way.* Logan, UT: Spendlove Medical Research Institute/2020 Program for Global Health/Profundities LLC. This study outlines a detailed comparison of the process of manufacturing an automobile tire with those of a pernicious, disease-causing virus, the Bluetongue virus. Process models are partially created and genomic/cladistic analysis is computerized and demonstrated using expressive knowledge tools with respect to automobile tires and the virus. Such analysis provides sound scientific groundings for organizational activity and supports permanent, beneficial outcomes.

achievement that we can enjoy, one that will forever enhance the prospects for those who follow in our footsteps.

You are not alone

Generalists are typically "heat-seeking" missiles. From the perspective of specialists, they could be considered "fair weather friends". They should be.

A valid generalist would immediately shift alliances once new developments have been found and validated. Indeed, a good generalist would often be the one to do the validation. In a fluid environment, generalists are involved in the process of designing the system, making use of fluidity and expressive knowledge forms. They would not be position to dictate outcomes in particular cases. Generalists would be very active in the adjudication and establishment of interpretive structures within the system, the keys and codes in particular that guide users through the processes in question.

To accomplish such tasks, generalists will tend to spread their influence, knowledge, and acquaintance throughout the scientific community. They will need to present the case for such new structures persuasively and judiciously. For one thing, as advocates for the new, they are charged with the introduction of new developments to general use. For another, their proposals may result in the invalidation of entire branches of study, reducing or eliminating the life's work of specialists with the introduction of new root cause elements.

For these reasons, generalists may or may not be popular. They should probably not be the kind of people that seek out popularity. They shouldn't be shy or retiring, however. Generalists need to be the kind of people that can comfortably work among others, establishing relationships and friendships, and forming alliances when necessary in support of valid developments.

Effective generalists need to be able to evangelize in support of root causes, to be persuasive as well as fair.

Root causes, virtue, and determination

Truly, virtue is not a direct component of any profession, although all require integrity and responsibility to some degree. Given the centrality of the generalist task, particularly in an environment in which generalists have direct control of important aspects of systems, it is essential to consider issues of trust and to arrange for limitations and controls to protect the integrity of the system should poor behavior surface.

Science for science's sake is an expensive proposition.

It is entirely possible to be engaged in scientific study and not be committed to understand and to manage root causes. In fact, from a purely scientific standpoint, that may be an uninteresting and counterproductive vocation. Looking for root causes may serve to skew the data, to encourage imbalanced education.

In a data-driven evaluative environment, the generalist has a most interesting task. In the absence of a health complaint, for instance, generalists have the opportunity to weigh in on health as a whole based on a comprehensive set of relationships. In order to accomplish this, such relationships need to be qualified according to accepted classification schemes. It is only through classification that the very complex can be understood, brought under control, and utilized.

The search for root causes can seem an arcane if not unpopular pastime. Institutionalizing such a pursuit can be considered one of the most fundamental achievements of our era. Why is this the case? Simply-stated, it is because discovery and support of one root cause can wipe out a multitude of money-making opportunities, at least in the current environment. Indeed, this is the whole point of the 2020 Program, the extermination of any and all existing diseases, *particularly* of the chronic kind.

This factor may at first present itself as being "anti-capitalist" if not "anti-market", but this is in no way true. The perception exists due to a simple misunderstanding. As Adam Smith wrote, capital commitments serve the public as they support trade in goods and services. Goods are of course good, the more the better. Many medical strategies are based on removing the bad. This is a beneficial thing to be sure, but it does not share the "more the merrier" condition. There is nothing *good* about disease, or disability, or pain, or anything of that sort. They should be relentlessly pursued and eliminated.[22] This is at the core of the "root cause" proposition.

The key is for generalists to help point the way to other, more legitimate commercial pursuits. In this, they need support. Much of this need come from government, which is ultimately responsible for establishing and supporting dual control. It may be difficult for many governments to serve in this capacity, as existing enterprises and their supporters tend to be powerful politically and economically. They may not be pleased, to be sure, with any societal "murmuration" leading away from their products and services.

Once such a system is in place, it takes no imagination to predict that individuals and interests benefiting from current conditions will focus their attentions on system participants, generalists in particular. Their will be enticements and threats. The "root cause" provision will clearly come under attack. How can this be anticipated and accounted for? This will require special care in the selection of participants, in providing competition among participants, and focusing their attentions only in their specific areas of interest.

Rigorous, appropriate selection process

How should generalists be selected? As in all things, the best recommendation should come from the literature. The best

---

22 Farnes, et al., *Optimum Performance Living*; Ostojic, D., Farnes, L., Lee, Sl, Tingey, , and Manicki, M. 2014. *The full enjoyment economy ... Mastering growth, maturity, and decline.* Logan, UT/Warsaw, Poland: 2020 Program for Global Health.

generalists will have been published as such. They will tend to be high in "centrality", having earned their positions and established their reputations by means of their prior works. In the scientific community, there are fundamental checks and balances. Scientists are subject to scrutiny at every level.

The first and best option is to make use of network science models and tools. Social Network Analysis has come a long way in providing the means of understanding and encouraging existing networks of trust and competence.[23] As a factor in the process, it would be wise to include publications and academic credentials in their profiles, including any patents and potential economic entanglements, as well as commercial affiliations.

To be published, their work must pass peer review processes, where other scientists review and approve the work. The tradition is that this be a blind review, although most would agree that such reviews are typically transparent, as the work of each researcher or team carries identifiers or watermarks that tip off peers with any acquaintance with the literature. Nonetheless, peer review does constitute a step in the process of establishing scientific knowledge.

The next form of "vote" that is easy to measure is the pattern of citations by other researchers. The fact is, if a work is important, subsequent papers on the subject will have to cite them if their work is to be seriously considered. Each citation can be considered a vote in a general sense, but in fact, the referring article may actually be criticizing the prior work, so more than a mere count of citations is not sufficient to establish "centrality" or acceptance.

Another factor is the strength of the referring article and of that researcher or team. Measuring such a phenomenon is the first concept that distinguished Google from other search engines

23 Tingey, K. B. 2009. Chapter Six: Complexity science and networks. In K. B. Tingey, *Methods-based management.* San Diego, CA: University Readers, 103-115.

when it came on the scene, the "PageRank" model at the base of the primary Google patent.[24] The old Yahoo searches with millions of poorly-sorted results immediately went by the board when Google's superior, citation-ranked approach became available.

The process can get all too cozy, however. Much criticism of the self-citation approach has come from within the scientific community and from without. Kuhn famously said that correction of the problem only comes through revolutions of a scientific nature.[25] Interestingly and uncomfortably, physical scientists who consider themselves more "scientific" than social scientists have been criticized by the social science camp for "socially constructing" their work.[26] This is criticism with a sting to be sure, coming from hard science's younger sibling, social science, which has obvious issues of its own.

Nonetheless, individuals that have distinguished themselves in their fields are likely to continue to distinguish themselves. To the degree that we are worried about competence, identifying them based on the record of their publications is a good thing. Knowledgeable, data-driven selection of generalists brings multiple rewards in the process of establishing a dual control environment.

Centrality as used in network analysis can be based on any of a number of factors. Assigning meaning to those factors is a critical aspect in understanding the population of generalists in question. For Google, it was simple: Which pages linked to your page,

---

24 Patent 6,285,999 by Lawrence Page. Key social network analysis concepts cited here were from Mizruchi, M. S., Mariolis, P., Schwartz, M., & Mintz, B. 1986. Techniques for disaggregating centrality scores in social networks. *Sociological methodology*, 26-48.
25 Kuhn, T.S. 1962. *The structure of scientific revolutions.* Chicago: University of Chicago Press.
26 Lincoln, Y. S., and Guba, E. G. 2000. Paradigmatic controversies, contradictions, and emerging confluences. In N K. Denzin and Y. S. Lincoln, *The handbook of qualitative research.* Thousand Oaks, CA: Sage Publications, 163-188.

ordered by the centrality of the ones that linked to them. Why did the links occur? This is a more nuanced kind of problem when you consider interactions among people and organizations.

Kenneth Tingey recounts an experience that sheds light on the issue.

> I carried out a social network analysis of federal government-sponsored, state-employed service providers with a colleague. These were people responsible for meeting the needs of people troubled with physical, cognitive, and emotional challenges. Their job was to work through the applicable science in each case, the policies in question, and available options in the local community to make their clients as self-sufficient and independent as possible. Undaunted by the well-known statistic that their collective efforts yielded a success rate nationwide of .5%, they carried on with something approximating confidence, mostly arguing for larger budgets, to meet the needs of their clients, of course.
>
> Excuse me if I seem more than a little jaded in my description of that group, but I bring the situation up as an example of the kind of challenge faced when establishing a dual control environment, particularly in the selection, preparation, and support of generalists, who serve in such an important capacity in such matters.
>
> In our study of the dynamics of the state agency, we simply asked the employees to indicate who they considered to be experts. This is to say, who did they go to for resolution of difficult cases. In the process, we got a pretty comprehensive response, not all, but most. The process was carried out in a confidential manner.

We "ran the numbers" on centrality and about eight individuals stood out. This was about 5% of the total. We met for a day with these eight people, sponsored by the agency. At this stage, the results had been announced, and participation in the session was clearly a "perk".

What we found out from the session was that there were two tiers of "experts". The top three were the three regional managers, late in their careers. Four others were young and mid-career professionals. The other expert did not fit either definition and was clearly not well-liked, but was known to be a confidant of the agency director.

We kept a transcript of the meeting. I evaluated and compared the statements of each person, keeping in mind that person's position in the social network. There was a clear division between the three older managers and the four younger experts. In short, the managers had been deemed experts because they knew of solutions that would not later be questioned. Such actions had nothing to do with science. In fact, they turned a blind eye to science as a valid input. It was all policy and tradition.

The younger experts were concerned with the science, they were more persuasive, more prepared to be so, and very careful to speak in indirect ways so as to not confront the managers. Their statements were high nuanced, with multiple levels of meaning. It was clear that there was a great deal of tension between the two groups. In fact, within the next six months, two of the younger experts, the ones with actual "expertise", had left the agency. Each contacted me after the project was over to tell me of their departures.

In one seminal moment, when we described an environment such as is proposed in the 2020 Program for Global Health, characterized by fluidity and dual control, one of the older managers threw his hands in the air and stated (paraphrased), "I know what you are talking about, I agree that it is a good thing and exactly what we need. I just want to be long gone when it happens."

Of course, in this, we are considering interactions among consumers of scientific knowledge, not the creators of it, but the same issues surely apply. As to an understanding of centrality, we had found two underlying definitions. First was a power-oriented construct, not really a cognitive sense of expertise. People went to the bosses for answers to questions because the bosses themselves would be the ones assessing performance. Having been with the agency for a long time, they "knew where the bodies were buried" and they were in position to reward those who followed them and to punish non-compliant workers. Due to their geographic spread throughout the state, such managers were thus able to function as "little autocrats" with the power to control all inputs and outputs, rewards and punishments.

Well, they were not able to control all. The mid-career workers constituted an alternate power structure, one that was probably more legitimate than the first, at least cognitively so. They were diligent in reviewing scientifically-grounded information as it was made available. They presented themselves with far wider vocabularies of underlying issues. They would be much more likely to thrive in a knowledge-driven environment.

Generalists serve key roles in establishing higher legitimacy where science comes together with society. In short, a dual control environment serves to transfer power from "regional autocrats" as just described and toward generalists and specialists themselves. Where a context presents itself in the system, regardless of geography and personality, it is the solutions documented within the system by the generalist and specialist experts that hold sway, applied by on-site professionals with respect for such knowledge and skill in working with the individuals in question, helping them to carry out such plans.

Competition among generalists

Competition is generally a good thing. In many cases, generalists with different perspectives could be charged to design models in parallel one with another. Data can serve to sort out the differences.

Administrators and policymakers can do much to encourage such competition. Indeed, fluidity and increased system legitimacy provides an improved policy environment, capable of clarifying underlying processes and validating outcomes by means of data collected in the process.

Much could be said about this phenomenon – of critical importance. At this point, we defer to documentation on the methods-based approach as to how competitive capacities can be both structured and supported.

Limitation of context

There is a tendency where there is not a conclusive system to support fluidity and adjudicate context for individuals who prove themselves and establish their expertise in one area to extend their influence to other areas. In part this is due to their heightened power and control over resources. In part it can be attributed to a generalized assumption that they are "smart". "Smartness" is thus assumed to be a fungible property. There may be some truth to this, at least in terms of general habits. Indeed, we are considering herein the phenomenon of the generalist, who is someone with a general level of knowledge and expertise.

This is a highly suspect assumption, however. In fact, if we follow it to its logical conclusion, it can lead us into a lot of trouble. It leads to the kind of thinking that provides justification to autocrats and bullies.

Robert Glaser and Mitchelene Chi, in their work on expertise, emphasize that know-how is centered in specific domains or areas of knowledge.[27] This is underscored by the writings of Dorothy Leonard and Walter Swapp.[28] Deep smarts as they call it requires decades of effort, repetitive, active, and consistent. There is ongoing debate on the subject, Malcolm Gladwell with his 10,000 hours of deliberate practice on one side[29] and a group that promotes a broader view of expertise and the requirements of achieving it.[30]

The point of our argument here is that influence within a system of dual control should be limited to certain contexts, partly from an expertise perspective, partly for balance and limited influence on system features.

27 Glaser, R. T., and Chi., M. T. H. 1998. "Overview"in M. T. H. Chi, R. Glaser, and M. J. Farr (Eds.), *The nature of expertise.* Hillsdale, NJ: Lawrence Erlbaum Associates, Publishers.
28 Leonard-Barton, D., Swap, W. C. 2005. Deep smarts: *How to cultivate and transfer enduring business wisdom.* Cambridge, MA: Harvard Business School Press.
29 Gladwell, M. 2011. *Outliers: The story of success.* New York: Little, Brown and Company.
30 Macnamara, B. N., Hambrick, D. Z., Oswald, F. L 2014. Deliberate practice and performance in music, games, sports, education, and professions: A meta-analysis. *Psychological Science, 25*(8):1608-1618.

Context and the trustworthy generalist

An environment characterized by fluidity does not feature a "sandbox" of any kind, where users, or even administrators, can "nose around" looking for things. With each branch of the trees in question, situations are progressively more defined, more specific with regard to context, continually restrictive with regard to anything that does not match the task at hand.

Society is based on trust. Conversely, politics is not. Well, it is to a degree, but without ongoing means of channeling that trust, it wears thin, sometimes descending into suspicion, hostility, and dissension.

Trust is the basis for legitimacy within science, but the scientific process itself is quite the opposite. Nothing is to be taken for granted. Hypothetical situations are defined in great detail, data acquired with great care, and interpretive methods are documented in great detail. The idea is that any claims made are to be replicated by independent groups, even competing groups, before they are taken seriously. Statistical methods are applied to rule out chance as a factor. In medicine, various trial runs of the idea are carried out and documented to be sure.

Once an idea passes through these gateways, it is believed to represent a natural phenomenon and can be relied on. In reality, however, seldom do situations present themselves neatly, as they were controlled in the scientific efforts. Furthermore, the same expansionary assumptions with regard to intelligence and expertise present themselves with regard to scientific findings. On their own, in the minds of the people and in application, they expand on themselves; they appear to apply to a wider context than had originally been established. This can be seen with interventions, pharmaceuticals, etc. Prerogative lurks at the basis of such activities. Since the treatment in question is allowed generally, and "since no one else is on the scene", it is tried for other uses. This is particularly true where there is money to be made, as outsized gain offers powerful enticements.

This is where our generalists come in. In the abstract, based on their knowledge and commitments to beneficial outcomes, generalists knit knowledge together in useful forms, linking the work of specialists and presenting results in useful forms. They benefit from expressive knowledge models, which help them tie things together in meaningful ways and to channel the activities of all in useful ways.

As we have discussed, all ideas are not created equal. With regard to the health and well-being of the people, ideas that capture root causes, whether they involve high performance or protection from disease, trump those that are related to downstream conditions. The logic of this is clear, although persistent success in this is very difficult. This is where generalists help us the most, however, by riveting our attention and directing our resources toward solutions and sound and lifestyles that are rich and fulfilling.

Alphabetical Index

45

47

49

www.ingramcontent.com/pod-product-compliance
Lightning Source LLC
Chambersburg PA
CBHW051820170526
45167CB00005B/2086